SIMPLE MACHINES ALL AROUND US

PULLEYS AT WORK

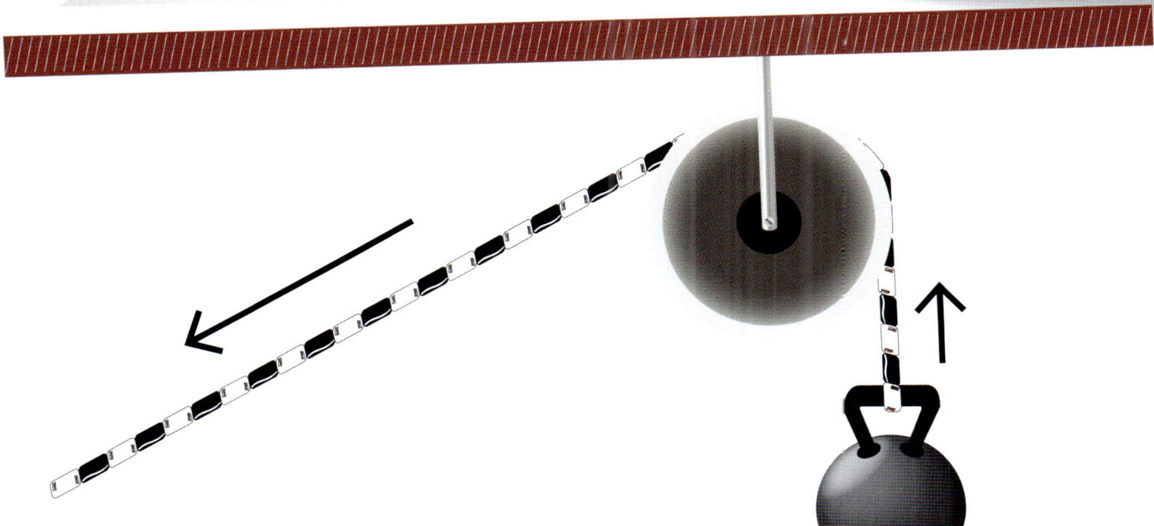

By Maggie S. Wesley

Enslow Publishing

DISCOVER!

Please visit our website, www.enslow.com. For a free color catalog of all our high-quality books, call toll free 1-800-398-2504 or fax 1-877-980-4454.

Library of Congress Cataloging-in-Publication Data

Names: Wesley, Maggie S., author.
Title: Pulleys at work / Maggie S. Wesley.
Description: New York : Enslow Publishing, [2022] | Series: Simple machines all Around us | Includes index.
Identifiers: LCCN 2020028565 (print) | LCCN 2020028566 (ebook) | ISBN 9781978520882 (library binding) | ISBN 9781978520868 (paperback) | ISBN 9781978520875 (set) | ISBN 9781978520899 (ebook)
Subjects: LCSH: Pulleys–Juvenile literature.
Classification: LCC TJ1103 .W47 2022 (print) | LCC TJ1103 (ebook) | DDC 621.8–dc23
LC record available at https://lccn.loc.gov/2020028565
LC ebook record available at https://lccn.loc.gov/2020028566

Published in 2022 by
Enslow Publishing
101 West 23rd Street, Suite #240
New York, NY 10011

Designer: Katelyn E. Reynolds
Editor: Kristen Nelson

Photo credits: Cover, pp. 1 ttsz/ iStock / Getty Images Plus; cover, pp. 1–24 (background and typeface) Reytr/Shutterstock.com; cover, pp. 1–24 (paper) jannoon028/Shutterstock.com; cover, pp. 1–24 (note sticker) Kindlena/Shutterstock.com; p 4 HAYKIRDI / iStock / Getty Images Plus; p. 5 FUTURE LIGHT/ Photolibrary / Getty Images Plus; pp. 7, 8 VectorMine/ iStock / Getty Images Plus; p. 9 Erik Isakson/Getty Images; p. 10 Anetlanda/ iStock / Getty Images Plus; p. 11 Tetra Images/Getty Images; p. 13 Mike Tauber/Getty Images; p. 15 Jenny P / FOAP/Getty Images; p. 17 Bettmann/Getty Images; p. 18 luplupme/ iStock / Getty Images Plus; p. 19 James Porter/ The Image Bank/Getty Images; p. 21 (bicycle gears and chain) Gonzalo Arroyo Moreno/Getty Images; p. 21 (treadmill) izusek/E+/Getty Images; p. 21 (bulldozer wheels) Philipp Berezhnoy/ iStock / Getty Images Plus; p. 21 (garage door opener) Smith Collection/Gado/Getty Images; p. 21 (crane) Jan Hakan Dahlstrom/Stone/Getty Images; p. 21 (window blinds) brizmaker/ iStock / Getty Images Plus.

Portions of this work were originally authored by Gillian Houghton Gosman and published as *Pulleys in Action*. All new material this edition authored by Maggie S. Wesley.

Printed in the United States of America

CPSIA compliance information: Batch #CSENS22: For further information contact Enslow Publishing, New York, New York, at 1-800-398-2504.

Contents

Boldface words appear in Words to Know.

The Important Pulley

Elevators help us reach buildings' highest floors. They wouldn't be possible without the pulley! A pulley is a simple machine, or an object with few parts that makes work easier. More than one pulley can help people lift heavy objects with less force.

AN ELEVATOR IS AN EXAMPLE OF A **COMPOUND MACHINE** THAT USES A PULLEY.

A Pulley's Pieces

There are several kinds of pulleys. Most have a wheel with a groove, or path, in it. A rope or **chain** runs along the groove. A pulley's wheel spins around an object called an axle or rod.

THERE ARE THREE KINDS OF PULLEYS: FIXED, MOVABLE, AND COMPOUND.

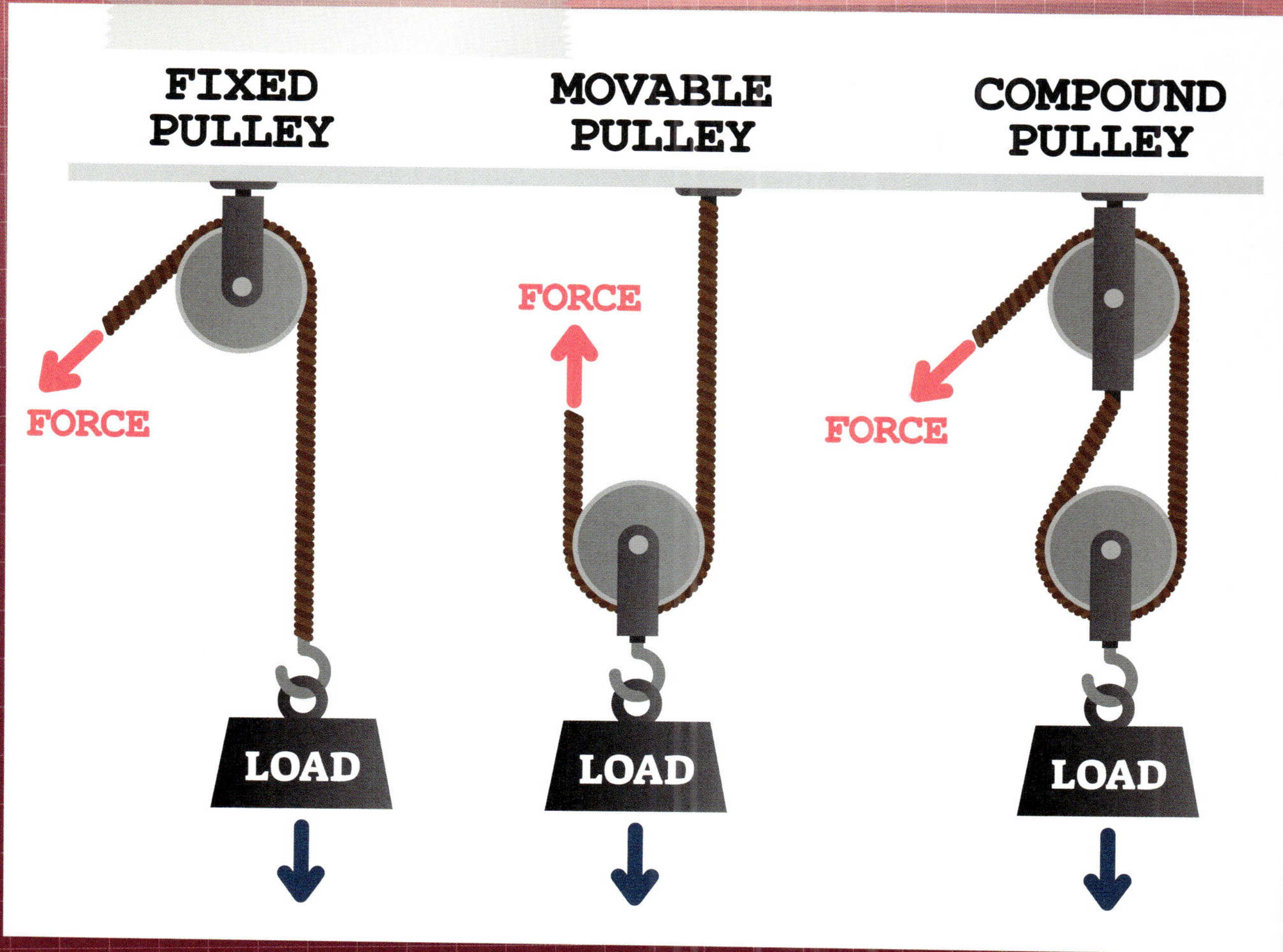

Figuring Out a Pulley's Forces

Pulleys can change certain forces. This is thanks to the wheel. The wheel allows the force used to pull down on one end of a rope to be reversed, or turned around. This lifts an object at the other end.

ADDING WHEELS TO A PULLEY RAISES THE PULLEY'S **MECHANICAL ADVANTAGE.**

The Fascinating Fixed Pulley

Fixed pulleys have an axle that's fixed in one place. When you pull down on one end of the rope, fixed pulleys redirect the force you use. They don't offer a mechanical advantage. However, they can be helpful for lifting objects overhead!

FLAGPOLES USE FIXED PULLEYS. YOU PULL THE FLAG UP BY PULLING DOWN ON A ROPE.

The Amazing Movable Pulley

A movable pulley moves along the rope with the load being moved. One end of the rope is attached, or bound, to something **stationary**. This object takes on some of the force to lift an object attached to the pulley's wheel, making it easier to lift.

ZIP LINES WORK USING A SINGLE MOVABLE PULLEY'S MECHANICAL ADVANTAGE.

THE COOL COMPOUND PULLEY

Compound pulleys use both fixed and movable pulleys, with two wheels making it easier to lift loads. Although you have to pull down farther on one end of a rope to make the work easier, compound pulleys still offer a mechanical advantage.

A COMPOUND PULLEY **SYSTEM** USED ON MANY SAILBOATS IS CALLED A BLOCK AND TACKLE.

The Pulley's History

People in Mesopotamia may have used the earliest pulleys to raise water around 1500 BCE. Greek **mathematician** Archimedes invented the block and tackle for loading and unloading ships around the third century BCE. Archimedes may have even invented a pulley system for moving ships!

PULLEYS MAY HAVE BEEN USED TO WATER THE HANGING GARDENS OF BABYLON AROUND 600 BCE!

Pulleys in Today's Machines

Pulleys can be found in many of today's useful compound machines. When a car won't run, you can call a tow truck. Tow trucks have a **hook** that connects to the car and a pulley system that makes moving the car much easier.

TOW TRUCKS OFTEN USE ANOTHER SIMPLE MACHINE, A LEVER, THAT STRENGTHENS AN APPLIED FORCE.

Find Pulleys Around You

Pulleys are more common than you may know. They may even be hidden! Look inside a car or truck and you'll find a part called the timing belt. It's a pulley that helps the **engine** run. Pulleys, like other simple machines, keep our lives moving!

HOW MANY OF THESE PULLEY SYSTEMS CAN YOU FIND NOW OR HAVE YOU SEEN THIS WEEK?

BICYCLE CHAIN AND GEARS

TREADMILL

BULLDOZER WHEELS

GARAGE DOOR OPENER

CRANE

WINDOW BLINDS

Words to Know

chain A series of metal rings joined together that can be used like a rope.

compound machine Two or more simple machines that work together to make a task or job easier.

engine A machine that makes power.

hook A tool that is curved or bent and is used for pulling, holding, or catching something.

mathematician Someone who studies math.

mechanical advantage The advantage, or benefit, gained by using a machine or tool to change a force.

stationary Something that stays in one place or position and doesn't move.

system A group of similar parts that move or work together.

For More Information

Books

Barger, Jeff. *Pulley*. North Mankato, MN: Rourke Educational Media, 2019.

Bethea, Nikole Brooks. *Making a Pulley System for Rapunzel*. Lake Elmo, MN: Focus Readers, 2020.

Crane, Cody. *Simple Machines*. North Mankato, MN: Children's Press, 2019.

Websites

Brain POP: Pulley
www.brainpop.com/technology/simplemachines/pulley/
Learn more about this simple machine and watch a video to see how it works!

Pulley
www.dkfindout.com/uk/science/simple-machines/pulleys/
This interactive site lets you explore the parts of a pulley.

What Is a Simple Machine?
www.wonderopolis.org/wonder/what-is-a-simple-machine
Have more of your questions about simple machines answered on this fun website.

Index